MACHINE TECHNOLOGY

Digging Machines

Amanda Earl and Danielle Sensier

Titles in the series:

Cutting Machines

Digging Machines

Mixing Machines

Spinning Machines

Cover inset: This mighty digging machine is called a backhoe loader. Find out more about how it works on pages 17 and 18.

Title page: A bucket wheel excavator is used for mining **material** from the surface. You can see how the machine works on pages 26 and 27.

Series and book editor: Geraldine Purcell
Series designer: Helen White
Series consultant: Barbara Shepherd, (former) LEA adviser on the Design and Technology National Curriculum.
Photo stylist: Zoë Hargreaves

First published in 1994 by Wayland (Publishers) Limited
61 Western Road, Hove, East Sussex BN3 1JD, England.
© Copyright 1994 Wayland (Publishers) Limited

British Library Cataloguing in Publication Data
Earl, Amanda
 Digging Machines. – (Machine Technology Series)
 I. Title II. Sensier, Danielle
 III. Bull, Peter IV. Series
 621.86

ISBN 0 7502 1278 0

DTP design by White Design
Printed and bound by L.E.G.O. S.p.A., Vicenza, Italy.

Words in **bold** appear in the glossary on page 30.

Contents

Spooning food

▲ The most simple machines we use are hand tools. A spoon is a simple digging tool. It is the right design for spooning flour from a bag into a mixing bowl. It is small enough to fit into the bag and the end is rounded so it can hold the flour – its **load**.

To pick up and move materials we sometimes need to use tools and machines to make the job easier.

The handle of a spoon is used as a **lever**. When the handle is pushed down the rounded end of the spoon is tipped up. This digging movement makes it easier to pick up the ice-cream this girl is eating.▼

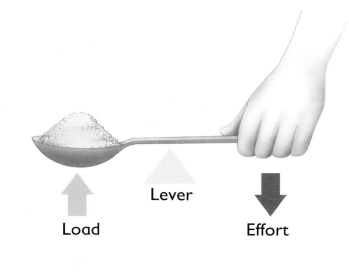

Load   Lever   Effort

Digging in sand

▲ A plastic spade is a simple digging tool. You have probably used one when digging in the sand. The children in the photograph above have filled buckets with dry sand to make sandcastles.

Digging in dry sand feels very different from digging in wet sand. Dry sand is soft and powdery, so it is easy to move. Wet sand is heavier and more solid, so you need to push the spade harder.

These children find that digging in wet sand needs more **effort** than digging in dry sand. They may need to use a spade of a different shape or size to dig in the wet sand.▼

Digging up soil

◀ Garden spades are used to dig up soil when working on an allotment. The soil needs to be turned over, ready for planting vegetables or flowers.

A garden spade has a long handle and a sharp cutting edge to slice into the soil.▼

Once the cutting edge has been pushed into the soil, the handle is tilted back and used as a lever. The point at which the spade moves backwards and forwards in the soil is called the **pivot**. The load of soil can then be lifted easily.▼

The handle of the spade is its lever.

Load

Effort

Because the spade's lever is long, it takes less effort to move a heavy load.

Pivot

Remember, the sharp edge of a spade can be dangerous, so always wear thick shoes for digging.

Ploughing fields

Ploughs are machines which farmers use to get the soil ready for planting new crops.

Early ploughs, like this horse-drawn plough, were very simple machines. They were made from a single, sharp wooden stick or blade pulled through the soil by a person or animal.▼

◀ Today, ploughs use a number of sharp metal blades which are v-shaped and curved. These blades are called ploughshares. They are pulled behind a tractor.

Each ploughshare slices through the soil and turns it over, making deep **furrows** in the soil.▼

This diagram shows how a ploughshare digs a furrow

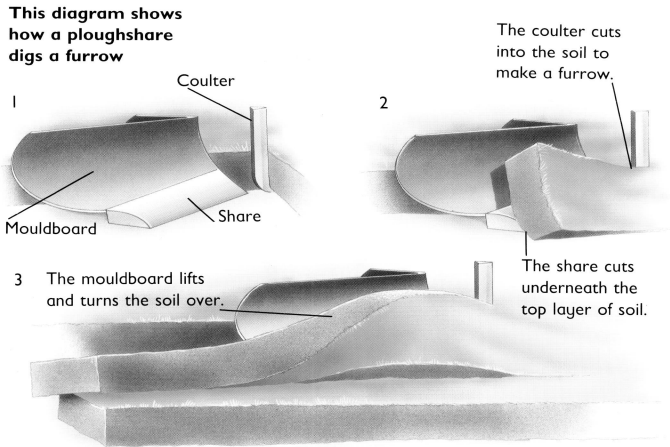

1

Coulter

Mouldboard

Share

2

The coulter cuts into the soil to make a furrow.

The share cuts underneath the top layer of soil.

3 The mouldboard lifts and turns the soil over.

Road breaking

Have you ever seen roadworkers digging up the road? Sometimes roads need to be repaired, or broken up ready for pipes to be laid down. This roadworker is using a hand-held digging machine called a road breaker.▼

A road breaker works like a hammer, hitting the road many times a second until the road surface is broken.

The road breaker is powered by a **compressor**, which pushes air in and out of the machine. The force of the air moves the sharp digging tool up and down.▼

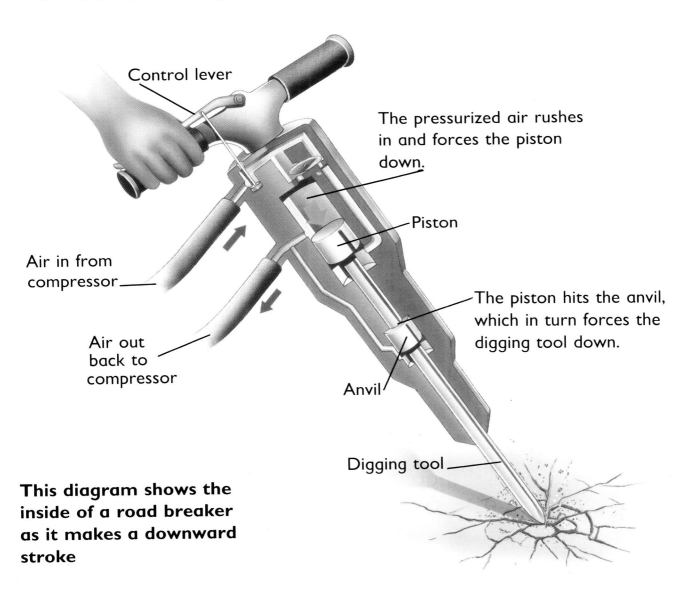

Control lever

The pressurized air rushes in and forces the piston down.

Air in from compressor

Piston

Air out back to compressor

The piston hits the anvil, which in turn forces the digging tool down.

Anvil

Digging tool

This diagram shows the inside of a road breaker as it makes a downward stroke

Building foundations

All buildings need strong bases to stand on, called foundations. These are built by digging deep holes in the ground, which are filled with **concrete**. The foundations stop the building from sinking or falling down.

Very tall buildings, such as skyscrapers, need deeper foundations. These are built in a different way from filling holes with concrete.

Instead, long blocks of steel or solid concrete, called piles, are hammered into the ground by a special digging machine. This machine is called a pile-driver. ▶

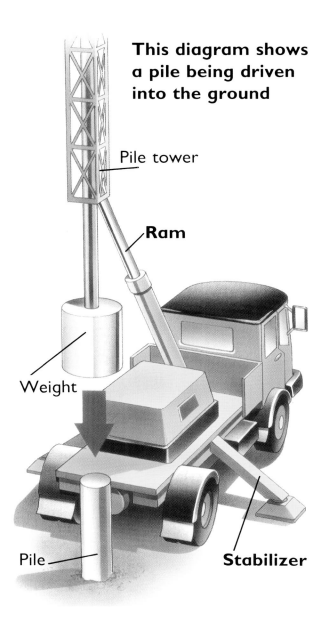

This diagram shows a pile being driven into the ground

Pile tower

Ram

Weight

Pile

Stabilizer

▲ A pile-driver has a large weight which is lifted and dropped on to each pile over and over again. This digging action forces the pile deep into the ground.

Excavating

Digging machines which dig and move earth or rubble are called **excavators**. They are often used on building sites and **quarries** to make work easier.

A mini excavator can be used when working in small spaces, like a garden. It has a single dipper arm, which acts as a lever to move the bucket in and out as it digs up soil.▼

▲ Larger excavators, such as this backhoe loader, can dig more than 500 spadefuls of earth at one time. It can do two kinds of work. At the back is a long arm and a bucket with sharp teeth, called a backhoe. This is used for digging up earth. The backhoe can move up and down or reach out a long way. At the front is a wider container called a loader. This is used for picking up the earth.

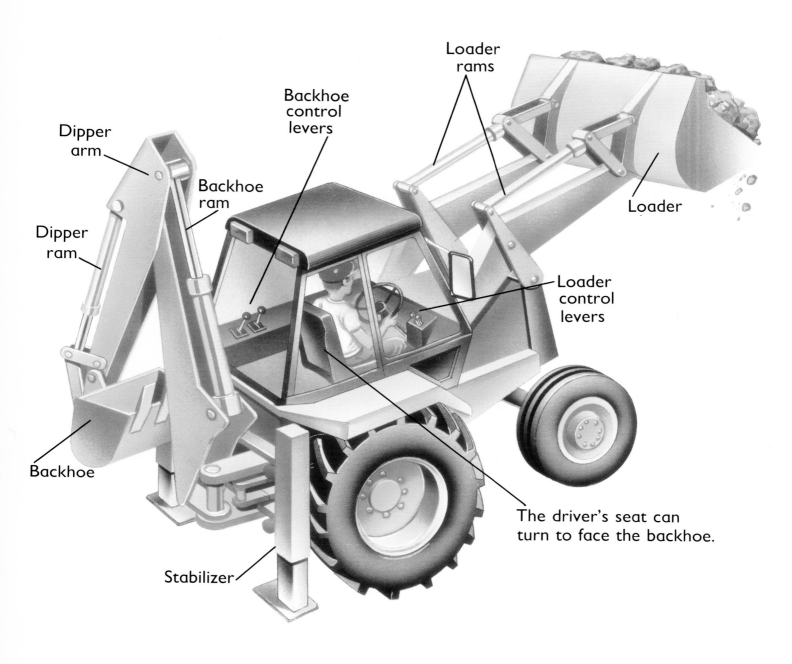

Dipper arm

Dipper ram

Backhoe ram

Backhoe control levers

Loader rams

Loader

Loader control levers

Backhoe

Stabilizer

The driver's seat can turn to face the backhoe.

▲ This diagram shows a backhoe loader at work lifting a load of rubble. There are many different kinds of excavators, designed to do different jobs. On the next page are two more examples.

▲ A split-bucket excavator has a large bucket which bites into the ground. It opens and closes, so that materials can be picked up and dropped more easily.

Notice how this backhoe digger has wide metal crawler tracks which fit over the wheels. These help it to climb up steep banks and stop it from sinking into soft muddy ground. ▶

Clearing ground

A bulldozer is a large digging machine, which clears and flattens rough ground. It is often used when building a new road.

The bulldozer's heavy metal blade moves earth and rocks out of the way. It is wide and curved, and made of solid steel. The blade is worked by four long arms which help it push and lift its load.▼

▲ At the back, there are metal claws, called rippers, to tear out boulders and tree stumps.

This digging machine needs a powerful engine to help it move heavy materials more easily. A tall pipe at the front takes away smoke from the engine.

Clearing snow

▲ When there is a heavy snowfall, powerful digging machines are needed. These are called snowploughs. They are used to clear deep snow from roads and airport runways.

A snowplough is usually attached to a tractor or a truck. It has a clearing blade at the front, like a bulldozer, which pushes the snow out of the way and clears a path for vehicles to use the road.

Snow can also block railway tracks, stopping trains from running. To help clear the snow, some trains have a railway plough. This is a v-shaped **fender**, which is fitted to the front of the train. The fender is like one huge ploughshare (see page 11). It pushes through deep snow, moving it away to each side of the track.▼

Digging under water

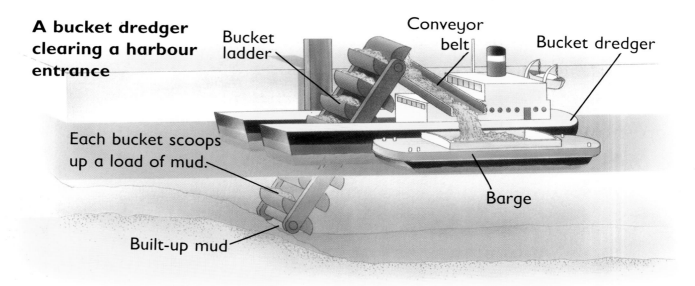

A bucket dredger clearing a harbour entrance

Bucket ladder

Conveyor belt

Bucket dredger

Each bucket scoops up a load of mud.

Barge

Built-up mud

Some digging machines work under water. Mud can block a harbour entrance or river and if it is not removed ships and boats cannot get through.

This is a floating digging machine, called a bucket dredger. It is used to dig up mud under the water. Can you see the row of buckets in the middle? These are on a chain which moves round and round. ▶

◀ The chain is called a ladder and it reaches down under the water. The buckets move along, digging into the mud and scooping it up. When each bucket reaches the top of the ladder, it tips its load down a **conveyor belt** into a waiting barge – a small boat used for carrying bulky loads.

▲ The grab dredger is another kind of floating digger. It is used in smaller spaces, like a **canal**.

It has one large split-bucket which can bite into the mud and weeds. The split-bucket is then lifted and the load is emptied into a barge, which takes it away.

Mining

Digging is needed when mining materials, such as coal, tin or iron **ore** – even diamonds.

A bucket wheel excavator can be used to dig up coal which is just under the ground. It has a huge digging wheel, which can be up to twenty metres high. When the wheel moves round and round its many buckets scrape up the coal. The buckets tip their load on to a wide conveyor belt, which moves very quickly.▼

A bucket wheel excavator digging coal

Railway trucks

Conveyor belt

Crawler tracks

Coal

At the end of the conveyor belt, railway trucks collect the coal and take it away to be used in homes or factories.

▲ This bucket wheel excavator is digging iron ore. It can move thousands of tonnes of material a day.

Digging a tunnel

Tunnel diggers do the biggest digging jobs of all. They are used to **bore** huge tunnels through hills and mountains or deep under the ground. The tunnels are needed for trains and cars – or sometimes to carry water.

A tunnel boring machine

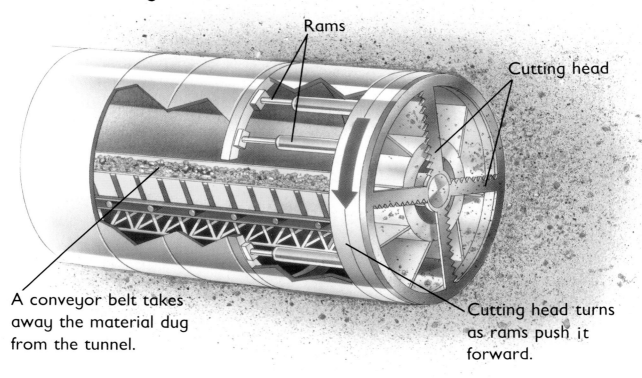

Rams

Cutting head

A conveyor belt takes away the material dug from the tunnel.

Cutting head turns as rams push it forward.

▲ The tunnel boring machine, known as a TBM, is up to eight metres wide and as long as a train. It is strong enough to cut through six metres of solid rock in just one hour.

▲ At the front of a TBM is a cutting head which spins round, and is pushed forward by powerful rams. It has hard metal cutters which tear through the earth and rock, digging out a long tunnel.

Glossary

Bore To dig a hole.

Canal A waterway which has been made to link two rivers or lakes.

Compressor A machine that makes pressurized air.

Concrete A building material made from water, sand, small stones and cement. Once it is dry it is solid as rock.

Conveyor belt A moving track to carry loads from one place to another.

Effort The force needed to make something work.

Excavators Machines which excavate (dig up or move) earth.

Fender A piece of metal which acts as a guard.

Furrows Long shallow trenches made by a plough.

Lever A simple lever is a bar or rod which tilts around a pivot. It can be pushed down at one end in order to raise or move a heavy object on the other end.

Load Something moved or carried by a machine.

Material Solid or loose matter, such as sand and earth.

Ore Rock that contains minerals or metals.

Pivot The point around which a lever turns easily.

Quarries Places where materials like stone or chalk are dug out of the ground.

Ram A rod which forces a machine or tool backwards or forwards.

Stabilizer A rod on a machine used to keep it steady.

Books to read

Diggers and Cranes (Flyers series) by Sally Hewitt & Nicola Wright (Colour Library Books, 1993)

Diggers and Dumpers (Eye Openers series) by Angela Royston (Dorling Kindersley, 1991)

The Kingfisher Book of How Things Work by Steve Parker (Kingfisher, 1991)

Lifting by Levers (How Things Work series) by Andrew Dunn (Wayland, 1991)

Machines (First Technology series) by John Williams (Wayland, 1993)

The Usborne Book of Diggers and Cranes by Caroline Young (Usborne, 1991)

Picture acknowledgements

Aspect Picture Library 10 (bottom) (K. Goff), 14 - 15 (A. Greeman); J. Allan Cash 10 - 11, 24 - 5; Cephas 25 (top) (F. B. Higham); Eye Ubiquitous 12 (Mostyn); Robert Harding Picture Library 19 (top) (M. Leslie Evans); Life File 16 (K. Curtis), 19 (bottom) (F. Smith), 21 (J. Fison); Q. A. Photos Ltd 29 (Eurotunnel); Tony Stone Worldwide *cover inset*, 17, 22 (G. Lincoln), 23 (P. Chesley); Wayland Picture Library/ (APM Studios/Photo stylist: Zoë Hargreaves) *cover background*, 4, 5, 6, 7, 8, 9/ *title page* and 27; ZEFA 20 (UWS). All artwork is by Peter Bull.

Index